MW00785650

Trolls

Jan Lindenberger with Joel Martone

Schiffer Publishing Ltd

4880 Lower Valley Road, Atglen, PA 19310 USA

Book Designed by : Randy L Hensley
Type set in: Text Korinna BT
Caps Korinna BT/Title Dom Bold BT

ISBN: 0-7643-0863-7

Printed in China
1 2 3 4

Published by Schiffer Publishing Ltd.
4880 Lower Valley Road
Atglen, PA 19310
Phone: (610) 593-1777; Fax: (610) 593-2002
E-mail: Schifferbk@aol.com
Please visit our web site catalog at
www.schifferbooks.com

This book may be purchased from the publisher.
Include $3.95 for shipping. Please try your bookstore first.
We are interested in hearing from authors
with book ideas on related subjects.
You may write for a free printed catalog.

In Europe, Schiffer books are distributed by
Bushwood Books
6 Marksbury Avenue
Kew Gardens
Surrey TW9 4JF England
Phone: 44 (0)181 392-8585; Fax: 44 (0)181 392-9876
E-mail: Bushwd@aol.com

Contents

Introduction

Welcome to the world of make believe and fantasy. When you let these little trolls, hobbits, and dwarfs into your life you will definitely expand your realm of imagination. Although they may only be constructed of vinyl, plastic, or rubber, they have a certain something that wins many of us over. They were designed for the child-at-heart and the fun-loving individual. Trolls live in trees, bushes, caves, burrows, and old hollow logs. They can be ugly or mischevious, but they are often grand cuties in the face. They may have long tails or pointed ears. How great to be an individual and show it! Trolls have been with us for centuries and will be here forever. Most true troll collectors do not put value ahead of appeal a troll has for them. Prices mean nothing; they cannot override your love for the wee folk. Even if you find one with no clothes or a stain, take him or her home and see your new friend smile.

Trolls were made by few manufacturers between the 1950s–1960s. Trolls originated in Scandinavian Nations of Norway, Denmark, Sweden and the Northern Islands. Scandinavian Dwarfs are often referred to as trolls, meaning evil spirits or monsters. They are also referred to as Hill People, because of where they were said to live. Trolls from Sweden, Denmark and Norway were known as thieves and non-heroes: ugly, cruel, and stupid. Others appeared as strong, magical, wealthy, and loving mischief. This book attempts to show you this variety of personalities, including trolls from the 1960s which had soft bodies with vinyl heads and hands. Modern trolls have a variety of shapes, sizes, hard and soft bodies, and body content (such as beans, beads, cotton, polyester). The Thomas Dam trolls have sexless bodies. The Dam troll banks are different, as the girls have rounded ears and the boys have sharply pointed ears. The 11", 12" and 3" sizes have the same bodies. Dam was the only company to use the same mold for the male and female body.

Thomas Dam trolls were made in the 1960s and marketed through Dam Things, Denmark, ITI, Thomas Dam Designs, Skandia House. It is possible that some of these were never marked. Dam introduced their trolls in 1959. They formed a partnership with Skandia House in 1966. After a court conflict, Skandia House gained exclusive rights for distributing and manufacturing in the United States. S.H.E.-marked trolls are actually Dam trolls. Trolls ran the marketing gamut in the United States until they became unpopular, but were still being produced for the European market. They returned to the United States in 1982 under the name of Norfin. Efs Marketing, Inc. had sole rights. Thomas Dam designed new creations until his death in 1989. The L. Kehm Company manufactured an unknown number of trolls, which were referred to as Space Alien, or monster trolls. They came in a variety of colors.

Russ Berrie started his business with $500 in a rented garage in 1963. His company has grown to over $250 million due to the fact he added more products to his line in 1992 (According to the company's annual report). Until 1991 he produced only the trolls, by the hundreds. The first trolls had the same body, head and hands but different outfits. In the 1960s he also manufactured Fuzzy Wuzzie, Bupkis Family, and Sillisculpts novelty statues.

Treasure Trolls/Ace Novelty Company produced the trolls in the 1990s with a special look—similar to the Russ and Norfin Trolls, only with a wishstone in its belly. The stone matched either the eyes or the hair or the costume. They came in shapes of diamonds, stars, circles, teardrops, rectangles and hearts. They were available in 4" and 8" hard bodies and a 12" soft body.

Norwegian Trolls were created in 1964 by artist Trygne Torgensen, in Tynset. 5 Nordic nations make trolls for export and tourist trade. Norwegians are known to describe trolls as ugly, made of wood and badly dressed. Many of these trolls are Vikings. A/S Nyform is one of the makers. They are made of one piece of plastic or rubber, and reflect the original Nordic concept of wicked and mischevious creatures. They have a variety of facial features including wide-open mouth with spaced teeth grins and large noses.

Uneeda Doll Company has been in business since 1917. In 1964 it began to produce the Wishnik line and has done so on and off for the last 30 years. Uneeda issued new Wishniks in sizes ranging from 2"–18" in the 1990s. They never did produce any of the novelty items. Unless the troll has a double horseshoe, Uneeda or Wishnik marks, it is not a Wishnik. In 1967 the Uneeda catalog noted the new "Way out" Wishniks and Moonatiks. Since they did not sell well they were discontinued. From the 1960s through the 1990s several series appeared on the market. They still come as the same design and in similar packaging as the early ones. There were relatively few companies manufacturing trolls in the 1950s and 1960s, but there were more than 40 known troll companies from the 1970s–1990s. Some marked their trolls, others didn't. The trolls may have painted-on eyes or paper eyes, but most trolls have inset eyes made of transparent plastic. The Uneeda Company referred to the Wishnik eyes as glassine. Most trolls' eyes are similar; the iris may be any color but the pupils are usually dark, amber being the most common. Gold and green are easily found. Blue eyes are rare. Some eyes appear to sparkle. Rhinestones are sometimes found on Wishniks.

PVC trolls are the most common. Trolls are also made from wood, plastic, plaster, metal, hard rubber, ceramic, fabric, to name a few. Their hair is made of mohair, yarn, rabbit fur, fake fur, thread, and nylon. Some even have molded hair. The more valuable ones are trolls with rooted hair. They are hard to find as they were more expensive to produce. They came as frozen, jointed, poseable, stuffed, bendable and clip-ons. Marks are found on the back of the neck or head, or the foot, or soles of feet. Dam marked their trolls with the date (The year created) and the company name. Other companies used tags and stickers which were usually thrown away by the owner.

Pricing trolls depends on condition, coloring, cleanliness, hair condition, cracked or missing eyes or clothes. Prices may also differ according to area found, from flea market to shop to show. One may not mind paying the price as it depends how badly one needs to fill their collection!

Happy Trolling!

Credits

A very special thanks to Joel Martone from Rhyme and Reason Antiques in Colorado Springs, Colorado. Joel packed up all his trolls (one of his many collections) and allowed me to take them to my studio to photograph them. With his help in pricing and his troll information this price guide was possible. Also thanks to 12-year-old Ashley Lindenberger (my granddaughter) for thinking ahead and collecting trolls for the past 8 years. This book is mainly from both Joel's and Ashley's vast collections.

Also thanks to the following people for their contribution of Trolls:

Carol Henderson of Divide, Colorado
Time In a Bottle Antiques at S. Virginia St. Antique Mall #3, Reno, Nevada
Carol Crane Antiques at S. Virginia St. Antique Mall #3, Reno, Nevada
The Brass Armadillo, Phoenix, Arizona
The Antique Center, Scottsdale, Arizona
The Antique Trove, Scottsdale, Arizona

Thank you for purchasing this Troll information and price guide. I hope you find it helpful as you go on your Trolling hunt. Remember that prices are not concrete as prices may differ according to location found as well as condition. Prices may also differ from shops to flea markets to garage sales. Again thanks and enjoy!

Jan Lindenberger

Bibliography

Peterson, Pat. Collector's Guide to Trolls. 1995: Collector Books, Paducah, Ky
Clark, Debra. Troll Identification and Price Guide. 1993: Hobby House Press, Inc., Cumberland, MD.

Dam and Norfin Trolls

Rubber movable head Pheasant with basket. 12". Dam
Things, "1964" printed on feet. $125–150

Rubber troll. Thomas Dam, 1977. 11". $50–75

Rubber with moveable head. Marked "Dam Things. Est.
1964" on foot. 12". $125–150

Rubber troll. 7". "Denmark, Foreign Pat. Pending"
on back. $50–75

Hard rubber trolls. Marked "Dam Things". 7". $75–100

Hard rubber troll with movable head. Marked "Thomas Dam". 5". $40–50

Hard rubber troll. Good Luck Mascot. Marked "Thomas Dam". $20–30

Rubber movable head troll. Dam Design. 5". $30–45

Hard Rubber movable head. Marked "Thomas Dam, Foreign Pat Pending". Denmark. 7". $125–150

Hard rubber clown troll. Thomas Dam. 7". $50–60

Rubber Caveman. "Thomas Dam" on neck. 5". $15–20

Rubber pirate troll. Dam. 7". $45–60

Hard rubber. Dam Things. 7". $100–125

Rubber Santa troll. 9" 1980. Dam, Denmark. "1980" on back. $50–60

Rubber girl troll. "Dam, made in China" on back. 8". $75–100

Rubber Santa troll. Thomas Dam,
Denmark. 1977. 14". $100–150

Rubber Mrs. Santa troll. Marked
"Thomas Dam, Denmark,
1977". 14". $100–150

Hard rubber lady troll. Marked "Dam Things." 1964. 12". $125–150

Rubber Santa troll. Marked "Thomas Dam, Denmark, 1977". 13" $100–150

Hard rubber Halloween troll. Marked "Thomas Dam, Denmark". 11". $75–100

Rubber troll with tail, dressed in burlap. 3". $65–85

Rubber boy troll. Marked "Dam". 7". $40–50

Rubber nude troll. Marked "Dam". 2 1/2". Missing felt cloths.
$15–20 dressed

Partially dressed rubber troll. Marked "Dam Things, 1964". 1".
$100–125 dressed

Rubber troll marked "Dam". 5". $40–50

Rubber nude troll. No marks. Dam. 2 1/2". $5–8

Rubber 2 1/2" girl troll. $7–10

Rubber girl troll. No marks. Dam. 2 1/2". $7–10

Rubber boy troll. Marked "Thomas Dam, made in Denmark". 5". $8–12

Rubber troll with movable head. Marked "Dam". 4 1/2". $15–20

Rubber troll with earring. Marked "Dam Design". 4 1/2". $15–20

Plastic transparent boy troll. Marked "Dam 1986". 5". $20–30

Top left: Plastic transparent nude troll. Marked "Dam 1986". 5". $20–30

Top right: Hard rubber bathing troll. Marked "Dam 1977". 12". $30–40

Left: Hard rubber girl troll. Marked "Dam". 5". $20–30

Bottom: Plastic 4 1/2" Trolls. Marked "Dam". $15–20 each

Rubber magician. Marked "Dam 1986". 4 1/2". $20–30

Hard rubber from Dam. Left top and bottom: Front and back of nude troll. Marked "made in China 1986". 5". $10–15

Rubber soccer fan troll. Marked "Dam 1986". 5". $20–25

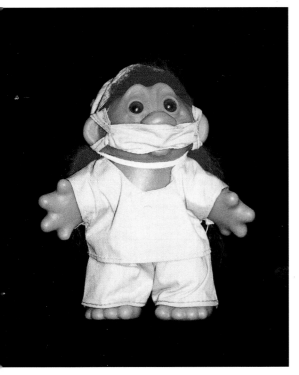

Rubber doctor troll. Marked "Dam 1986". 4 1/2". $20–25

Same troll, only with clothes. $45–65

Rubber troll with painted-on pants. (Shown nude) Marked "Dam 1980". 7". $45–65

Transparent plastic babies. Marked "Made in China". Dam. 3–3 1/2" $10–15 each

Plastic boy troll. Marked
"Dam 1984". 5". $20–25

Plastic girl troll. Dam. 3". $8–12

Rubber bunny troll. Dam. 2". $6–8

Rubber college troll. Marked "Dam Things 1964".
2.5". $15–20

18

Rubber Sea P' Troll. Marked "Thomas Dam Denmark 1977". Norfin tag. 9 3/4". $30–40

Rubber cheerleader troll. Marked "Dam Things". Hard to find. $20–30

Rubber girl troll. Marked "Thomas Dam". 9". $30–40

Rubber Astronorf Norfin Troll. 9 3/4". $30–40

19

Top left: Hard plastic Hanukkah Troll.
Dam. Norfin tag. 1986. 4". $15–20

Left: Rubber birthday girl troll. Betty. 4".
Norfin. $15–20

Top: Rubber girl troll. 3". Dam. $10–15

Bottom: Rubber trolls. Dam.
Norfin tags. 5". $10–15 each

Rubber trolls.
Dam. 5"–6".
$10–15 each

Rubber trolls. Dam.
Norfin tags. 5".
$10–15 each

Rubber baby and Libra trolls. Norfin tag. 3". Marked
"Made in Denmark 1986". $15–20

Rubber clown troll. Marked "Dam 1986
Denmark". 5". $15–20

21

Rubber infant in box. 3". $10–15

Little Tykes rubber Norfin Troll. Marked "Dam Design Denmark". 3". $10–15

Rubber baby. 3". $10–15

Rubber Norfin tagged trolls. 10". Marked "Dam 1982". $50–75 each

Rubber troll. Marked "Dam Denmark 1982". 11". $50–75

Rubber grooms. Marked "Dam 1985". 3". $15–20 each

Rubber Norfins. Marked "Korea 1985". 3". $10–15 each

Rubber Santa. Marked "Dam Design Denmark". 3". $10–15

Rubber little girl troll.
Marked "China". Dam.
3". $10–15

Rubber troll in suit with bunny ears. Movable head. Marked "Made in China". Dam Norfin 1984. 3". $20–25

Rubber little boy troll. Marked "China". Dam. 3". $10–15

Rubber boy troll in red suit. Marked "China". 1988. Dam. 3.5". $10–15

Rubber nude troll. "Bend–ems–Justoys". 1994. 4". $10–12

Fabric troll. Marked "Norfin Troll,
Justoys". 1992. 18". $15–25

Rubber Norfin monk. Marked
"Dam 1982". 11". $50–75

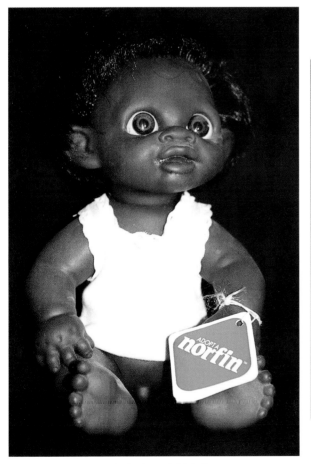

Rubber baby boy doll Troll.
Norfin. Jointed. 12" $75–100

Stuffed troll, nylon. Norfin. $20–30

Rubber Norfin clown. Marked "Dam Denmark 1988". 6". $20–30

Velour sailor girl troll. Norfin. 8". $5–8

Russ Trolls

Rubber Russ trolls. 10". $15–20

Rubber Russ trolls.
5". $10–15 each

Rubber Mr. and Mrs. Santa ornament trolls. 5". $10–15

Rubber Russ troll. 3". $10–15

Rubber Russ troll. 4". $10–15 each

Rubber Russ troll $6–9

Rubber Russ trolls. 4". $10–15 each

Rubber Russ trolls. 4". $10–15 each

Rubber Russ trolls. 4".
$10–15 each

Rubber
Russ troll.
4". $10–15

Rubber Russ troll. Marked
"China". 3". $10–15

Rubber Russ troll.
Team sticker on
foot. 4.5".$10–15

31

Rubber Russ troll. 1992. 5". $8–10

Rubber toy troll by Russ. Came from top
of candy holder. 3". $2–3 as is

Rubber Good Luck Pocket Troll by Russ. 1.5". $5–7

Hard rubber Christmas trolls. Russ. Hard to find. Being remade
today. 3". $20–25 old. $3–4 new

Rubber Russ trolls.
3". $6–8 each

Rubber Russ mini
figurines. 3". $5–7 each

Rubber figurines. Russ. 3". $7–10 each

Rubber Russ baby. Hard
to find. $7–10 each

Rubber Troll babies. Movable heads. 2". $5–7 each

Rubber Russ troll holding apple. Marked "Russ" and "China". 3". $10–15

Rubber Troll baby. Movable head. 2". $5–7

Hard rubber trolls. Marked "Russ, China". 5". $10–15 each

Rubber head plush "The Boss" troll with stuffed body. Russ. 8". $20–25

Rubber head, cloth body 49ers troll. From NFL Troll Kidz. 11.5". $35–45

Rubber head, cloth body Raiders troll. 11". $35–45

Padres troll with rubber head and cloth stuffed body . 11". $35–40

Rubber poseable trolls Tracey and Travis. 1993. Russ. 8". $20–25

Rubber head troll with velour and and stuffed body. 6". $10–15

Rubber poseable Tracey Troll. 1993. 0". $20–25

Rubber head troll with fleece and stuffed body. Russ. 8". $15–20

Rubber head Trolls with cloth stuffed body. 13.5". **Left to right**: Tootsie, Pebbles, Mrs. Santa, Spanky. 13.5". By Troll Kidz. $20–30 each

Rubber head troll ballerina with velour body. Russ. $20–30

Rubber head troll with cloth body. 7". Russ. $10–15

Rubber head trolls with cloth stuffed bodies. Russ. 13.5". By Troll Kidz. $20–30 each

37

Above and below:
Rubber head troll with cloth stuffed body. Wee Troll Kidz.
Russ. 12". $20–30

Rubber head cloth body undressed troll. Russ. Pookie. 12.5".
$10–15

Rubber head cloth body troll. Russ. 9". $15–20

Troll with rubber head and cloth body. Russ. 12.5". Pee Wee. $15–20

Jumbo stuffed clown troll. Rubber head satin stuffed body. Russ. 22.5". $30–45

Uneeda Wishnik Trolls

"Sock it to me" troll.(From the show *Laugh In)* Rubber with moveable head. 6". Wishnik. $60–75

Rubber nude Double Nik. 1967. $50–60

Rubber nude Double Nik. 1967. $50–60

Rubber Double Niks. Uneeda. 1965. $50–60 each

Rubber dark skin troll. Marked "Wishnik".
6.5". $30–40

Rubber troll. Marked "Wishnik, Uneeda". 5.5". $30–40

Rubber light skin trolls. Marked "Wishnik". 6.5". $30–40 each

Rubber Luv Nik. Missing hair. Marked "Wishnik, Uneeda".
Hong Kong. 5". $20–25

Rubber "Sock it to me" troll. Wishnik. 5.5"
$50–60

Rubber nude troll. Wishnik. 5.5". $15–20

Rubber nude troll. 5.5". Wishnik. $15–20

Rubber cowboy troll. Horseshoes on
feet. 3.5". $20–25

Rubber nude Wishniks. Marked "Uneeda Doll Co." 3". $5–8

Rubber nude Wishnik. Uneeda Doll Co. 3". $5–8

Rubber Wishnik nude troll. 3". $5–8

Rubber Wishnik. Uneeda. 1991. 3". $5–8

Rubber Wishniks. 3". $5–8 each

Rubber Wishnik troll. Push button in his belly and his eyes light up. 4". $10–15

Rubber Wishnik troll. Horseshoes on feet. 3". $10–15

Rubber Golf–nik. Bendable. Marked "Uneeda, Hong Kong". 5". $30–35 in the box

Rubber Pik–Nik. Bendable. Wishnik. Marked "Hong Kong". 5". $15–20 Others in this series are Cook–Nik, Quick–Nik, Luv–Nik, Rock–Nik. 5". $15–20 as is

Rubber graduate. Marked "Uneeda Co Inc, China". 8". $20–25

Rubber Golf–Nik, out of box. 5". $20–25

Rubber Cowboy. Marked "Wishnik by Uneeda". 8". $15–20.

Rubber graduate by Uneeda, Wishnik. 6". $20–30

Rubber trolls. Marked "Wishnik, Uneeda Doll Co., Inc." 8". $20–25

Rubber Farmer. Marked "Wishnik, Uneeda Doll Co Inc, China". 8". $20–25

Rubber Tennis and Genie Trolls. 8". $35–40 each in package

Rubber trolls in packages. Marked
"Wishnik, Uneeda Doll Co, Inc.
China". 8". $35–40 each

Rubber Batman. Marked Wishnik. 5". $50–60

Applause Trolls

Rubber "Magic Troll" babies by Applause. Pale color. Marked 1991. 3". $4–6. Not poseable. These babies each came in a package which contained magic stardust and a magic bag. By adding water to the stardust a charm appeared as the bag dissolved. With each troll came a bracelet for the charm.

Rubber "Magic Troll" babies by Applause. Marked 1991. Flesh color. 3". Not poseable. $4–6 each.

Plastic "Magic Troll" babies by Applause. Marked 1991. 3". Moveable arms and legs. $4–6 each

All rubber Troll dolls. Applause. Marked 1991. 11". $20–25 each

Rubber poseable Ollie Boy troll marked "Sreet Kids". Ollie Boy came packaged with a skateboard. There are four in this Total Troller series: Ollie Boy, Grinder, Thrasher and Airwalker. 6.5". $8–12

Troll baby on the right. Rubber head, arms, feet with cloth body. Applause. Marked 1991. 11". $20–30. Left troll baby is marked "1991 Applause" and has a plush body with rubber head. 11". $20–30

Rubber fully-jointed troll dolls by Applause. 5". $8–12 each

Ace Novelty Trolls

Rubber mini trolls. Marked "Ace Novelty Co; China". 1.5". $4–6 each

Rubber troll marked "Ace Novelty Co; China".
4.5". $6–8

Rubber girl troll with jewel in
belly. Marked "Ace Novelty Co;
Made in China". 4". $10–15

Rubber trolls marked "Ace Novelty;
Made in China". 4". $10–15 each

Rubber trolls marked "Ace Novelty; Made in China". 4".
$10–15 each

Rubber nude trolls with jewels in belly. Marked "Ace Novelty Co; Made in China". 4.5". $8–12 each

Rubber troll figures. Marked "Ace Novelty Co". 2". $5–7 each

Rubber Trolls. Marked Ace Novelty Co; Made in China.

Rubber girl troll with jewel in belly. Marked "Ace Novelty Co; Made in China". 4.5". $8–12

Left: Rubber head, cloth body. Marked "Ace Novelty Co; made in China". 12". $25–35

Center Left: Rubber head, cloth body with jewel in belly. Ace Novelty Co. 12". $25–40

Center Right: Rubber head, cloth body. Ace Novelty Co. 12". $25–40

Rubber trolls in box. Treasure Trolls. Ace Novelty Co; made in China. 4.5". $10–15 each

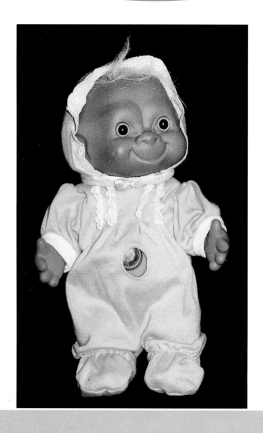

Rubber newborn baby. Treasure Troll. Ace Novelty, China. 12" Nude $15–20. Dressed $25–40

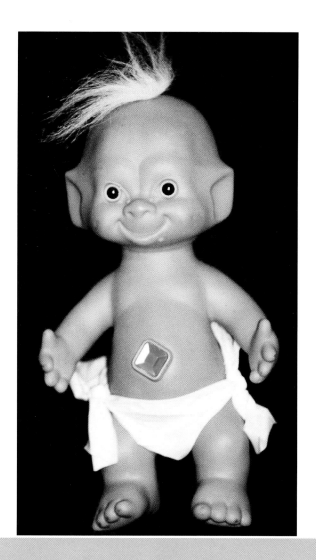

Norwegian and German Trolls

Rubber figurine by Heico. 7". $60–75

Rubber-bodied troll with tail. N.C. Ressler, Denmark. 3.5" tall. $40–60

Hard plastic girl troll. Came with a tag marked "A/S Nyform". 5.5". $60–75

Hard plastic figure of an old troll woman with a stick. Marked A/S Nyform. Tynset, Norway. 7". $60–75

Hard plastic Banjo player troll figure. Marked "Nyform, Norway". 6". $75–100

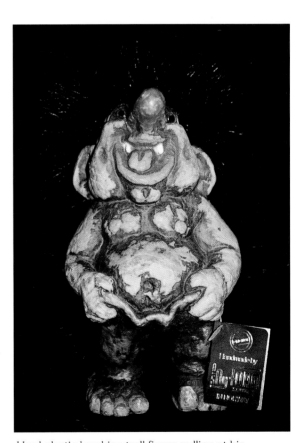

Hard plastic laughing troll figure pulling at his pants. Marked A/S Nyform, Norway. 6".$75–100

Hard plastic sitting troll figure. Marked "Nyform". 7". $75–100

Hard plastic man figure troll, with his hands in his pockets. Marked "A/S Nyform, Tynset, Norway". 5.5". $60–75

Hard plastic little girl troll figure. Marked "A/S Nyform, Norway". 7". $50–75

Hard plastic troll sitting on a stump. Marked "Nyform". 11". $60–75

Nyform troll with tail. Early. 5". $75–100

Latex skier troll. Marked "Nyform". 10". $125–150

Nyform troll with tail. Early. 10". $100–125

Wooden Viking troll. Marked "Denmark". 11".
$15–20

Hand-carved wooden girl troll. Henning, Norway. 3.5". $35–45

Wooden Viking troll with metal cap and chain. He held a stick in his outstretched hand. No marks. Norwegian. 9". $15–20

Hand-carved wooden troll. Henning, Norway. 6". $40–50

Wooden Viking troll. Japan. 2.5". $10–12

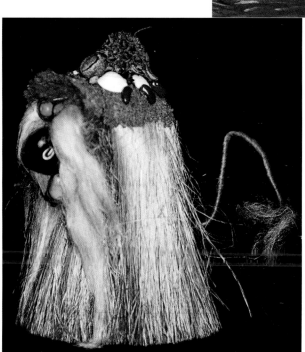

Wood and straw Viking with green eyes and a large brown nose. Rocky Mountain Troll. Marked Arensbak, 1978. 11.5". $35–50

Hand-carved wooden troll. Henning, Norway. 6". $40–50

Trolls Made in China

Hard rubber girl and boy trolls. Each girl came with boy in package with a set of rings to match. Made by M.T. China. $10–15 set

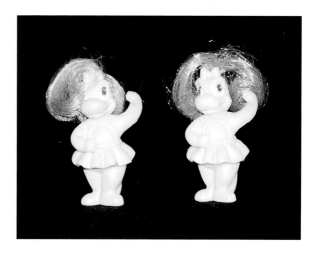

PVC dinosaurs figurines. Marked "M.T." China. 4". $8–10 each

Rubber head, cloth, stuffed
body. Made by Mul Ti Toys
Corp. China. 9". $15–20 each

Rubber baby with
molded diaper.
Marked "M.T."
China. 4". $10–15

Plastic Troll figure. His
clothes are molded with
body. Marked "M.T."
China. 3". $20–30

Rubber body builders. Marked "Soma". 2". $20–30 set

Rubber body
builder. Marked
"Soma". 2". $5–8

Rubber elf troll.
Marked "I.T.B."
China. 7". $20–25

Rubber head, cloth stuffed body troll dolls. Marked "Soma". 10". $10–15 each

Rubber ballerina troll.
Marked "I.T.B."
China. 7". $20–25

Rubber nude troll. Marked "1992,
I.T.B." China. 5". $15–20

Rubber head, cloth stuffed body troll. Marked
"I.T.B." China 1991. 12". $20–30

Rubber bathing beauty trolls. Marked "I.T.B." China. 5". $20–25

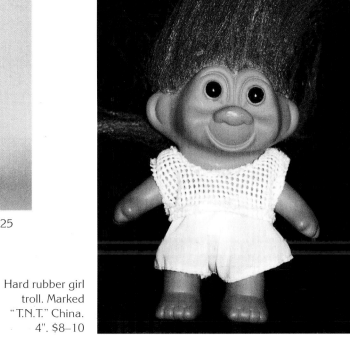

Hard rubber girl
troll. Marked
"T.N.T." China.
4". $8–10

Rubber trolls. Marked "T.N.T." China. 5". $8–10 each

Rubber Nasty Trolls, marked "Toys and Things. T.N.T 1991". Press belly and he sticks out his tongue. Missing tongue. $7–10

Rubber trolls. Marked "T.N.T." China. 5". $8–10 each

Rubber trolls. Marked "T.N.T." China. 4". $8–10 each

Rubber trolls. Marked "China". 4". $8–10 each

Rubber head plastic body girl troll. Her hair has been tied up. Made in China. 4". $15–20

Rubber sheriff troll. 3". Made in China. $8–10

Rubber chef troll. Made in China. 4". $8–10

Rubber trolls. Made in China. 4". $8–10

Rubber light-colored boy troll. China. 7". $15–20

Rubber boxer troll. Made in China. 2.5". $7–10

Rubber flesh-colored girl troll. China. 7". $15–20

Hard plastic nude troll. 3". China. $2–3

Rubber head, cloth stuffed body, talking clown troll. Press his body and he talks. China. 9". $20–30

Rubber head, stuffed body troll dolls. China. 11". $20–30 each

Rubber troll with sunglasses. China. 3.5". $5–8

Rubber mini
nude troll. China.
2". $3–5

Plastic head, stuffed body clown troll.
China. 8". $10–15

Rubber animal head
girl troll. Made in
China. 6". $10–15

Hard rubber ballerina troll. China. 7". $15–20

Miscellaneous Company Trolls

Rubber head troll with plastic body. Marked
"Marx". 2.5". $15–20

All-rubber gorilla. Movable head.
Hong Kong. 3.5". $20–25

Soft rubber "Lucky Lou Troll". Marked "Marx".
Hong Kong. $15–20

Rubber head troll with plastic body.
No marks. 6.5". Hong Kong. $15–20

Soft plastic winking eye troll. Left, 3". Right, 4". Hong Kong. $10–15 each

Soft rubber troll with fangs. 3". Marked "Hong Kong". $15–20

Rubber head troll with plastic movable arms and legs. 4". Hong Kong. $30–40

Sailor troll with binoculars. 3". Marked "Hong Kong". $8–12

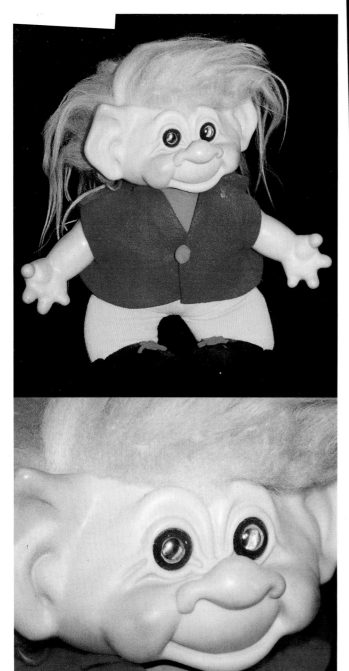

Troll with rubber head and arms, and stuffed body. 11". Scandia House Enterprises. Dark complexion. $100–125

Rubber head and arm troll with stuffed body. 11". Scandia House Enterprises. Light complexion. $100–125

Rubber "True Troll" by Scandia House. 1965. 2.5". $10–15

Leprechaun Troll with hard rubber head and arms, and corduroy body. Scandia House Enterprises. 11". Marked "S.H.E." $75–100

Satin stuffed Troll. 1992. 12". Dandee International. $10–15

Rubber collegiate troll, Yale. 2 3/4". $15–20

Hard rubber caveman troll. Marked "Neanderthal Man by Bijou Toys". 1963. $30–40

Plastic troll. Marked
"The Topps Co."
3". $8–12

Rubber blue Moon Monster. 1964.
3". L.Khem. $35–45

Rubber green Moon Monster. 1964. 3". L. Khem. $35–45

Rubber troll girl Moon Monster. 6".
Unmarked, Marx. $50–75 Rare

Rubber troll. 1964. L.Khem. 3". $35–45

Rubber nude baby trolls. 1982. 1.5".
Creata. $15–20 each

Rubber troll figure.
Creata. 3.5". $6–8

Trolls with rubber heads and cloth stuffed
bodies. **Left to right:** Pajama Baby, Nude Baby,
Talker baby. H.Y.I. China. 12". $20–30 each

Rubber cowboy troll. Creative
Manufacturing, Inc. 1978. 3". $20–25

Rubber troll with earring.
Bright of America. 4". $7–10

Rubber girl troll.
Bright of
America. Made in
China. 4". $7–10

Rubber hobbit troll. Royalty Industries Inc.
1973. 8". $50–75

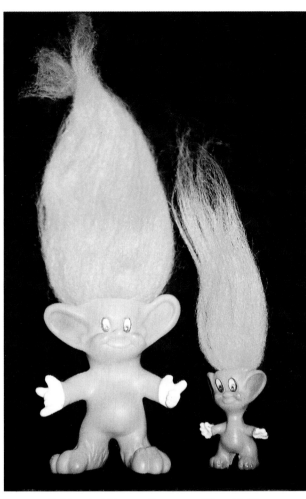

Rubber hobbit pencil toppers. 3" and 1.5".
Royalty Industries Inc. $15–20 each

Rubber troll with movable head, legs, arms.
Marked Unica, Belgium. 1965. 8". $40–50

Rubber bunny troll. Lucky Troll. Koalanme,
Hawaii. 1982. 4". $30–40

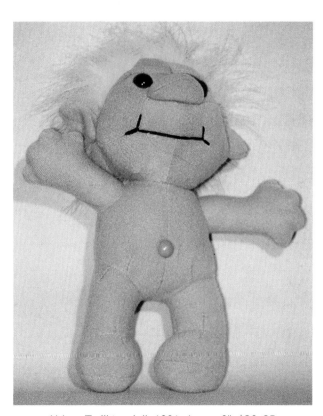

Velour Trollkins doll. 1991. Acme. 9". $20–25

Rubber clip-on trolls. Left to right: Russ, Korea, China. 3.5" to 3". $3–5 each

Plastic nude troll. Korea. 3.5". $4–5

Plastic Grandpa troll. Korea. 5.5". $8–12

Rubber action trolls. Hasbro. China. Rotates from the waist. 4". 1992. $5–8 each

Rubber action trolls. Hasbro. China. Rotates from the waist. 5". 1992. $5–8 each

Rubber action battle trolls. Hasbro. China. Rotates from the waist. 5". 1992. $5–8 each

Rubber head plastic body Action Troll by TNT. 1992. China. $6–10

Action trolls. Plastic movable arms and legs. 4". $6–10 each

Rubber Pirate Troll. Cap'n Troll. Missing spear.
Hasbro, 1992. China. $6–10

Complete set of 12 hard rubber trolls.
Imperial Toy Corp. 4". 1998. $5–8 each.

Rubber "Burger King Kids
Klub" Trolls. Glow in the
dark figures. Heads
rotate. 1993. Made in
China. 2.5". $5–7 each

Treasure Troll. This set came as a collector's first edition set of 12. 3". Galoob Toys. 1998. Included in this set are Leenie, Sukey, KiKi, Spritsy, Tansy, Bala, Lolly, Piny, Sunny, Trinkie, Floressa, Jan Jan. $20–25 each. Another set of 3" Trolls from the Galoob Company are Ele, Jo Jo, Poko, Tish, Ono, Tootsie, Hulu, Olani, Lyra, Effie, Zia, Zoa. $8–10 each

Plastic Fantasy Friends Troll. 7". Art Mark, Chicago. Korea. $4–6

Hard rubber Fantasy Friends magnets. 2.5". $2–3

Banks

Hard rubber bank of nude troll. Marked "Royalty Design of Fla." 1967. This hobo has lost his clothes. Originally had felt pants and shirt. Molded shoes. $50–60 mint

Rubber Thomas Dam troll bank. 7". $60–80

Rubber Thomas Dam troll bank. 7". $60–80

Rubber Troll bank. Marked "RFD 1968". 6.5". $60–80

Rubber Thomas Dam troll bank. 7". $60–80

Rubber troll pirate bank. 7". $65–85

Rubber Thomas Dam troll bank. Marked "Made in Denmark". 7". $75–100

Rubber troll bank. Thomas Dam. 7". $60–80

Rubber troll bank. Thomas Dam. 7". $60–80

Rubber troll bank. Thomas Dam. 7". $60–80

Rubber troll bank. Thomas Dam. 7". $60–80

Rubber troll bank. Thomas Dam. 7". $60–80

Rubber troll bank. Thomas Dam. 7". $60–80

Rubber troll bank. Thomas Dam. 7". $60–80

Rubber troll bank. Thomas Dam. 7". $60–80

Rubber troll bank. Thomas Dam. 7". $60–80

Rubber troll bank. Thomas Dam. 7". $60–80

Rubber Santa Troll bank. Marked
"Denmark". 7". $65–85

Hard rubber cowgirl troll bank with molded
clothes from Creative Manufacturing. 1978.
8.5". Rare. $75–100

Rubber Santa troll bank. Thomas Dam. 7". $100–125

Rubber warrior troll bank. Dam Things. 8". $50–75

Hard plastic seal bank, Norfin. "Thomas Dam 1984" marked on right flipper. "Denmark" on left flipper. 6.5". $50–75

Hard plastic bear bank, Norfin. "Thomas Dam" marked on foot. 7". $50–75

Plastic gumball bank machine by Ace Novelty. $15–20

Plastic "Total Trollers" bank by Just Kids of California. 15". Came with cookies in it. $10–15

Ceramic troll bank by Russ. $20–25

Rubber mouse troll bank. Royalty Design. 11". $40–50

Hard rubber "Street Kids" troll bank. 1981. 9". $25–35

Rubber Santa mouse troll bank. 10". $40–50

Rubber sleepy troll bank. Made in China by HEI. 1992. 6". $20–30

Nodders

Rubber troll nodder. "I love you this much".
Marked "Hong Kong Berries". 1972. 5". $45–65

Hard rubber troll nodder. "Happy Birthday".
Marked "Hong Kong Berries". 4". 1972. $50–75

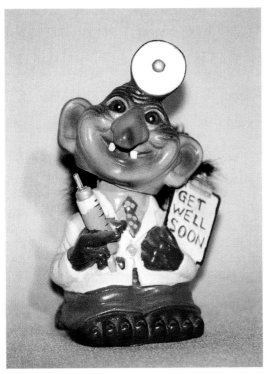

Hard rubber troll nodder. "Get well soon".
Marked "Hong Kong Berries". 4". 1972. $50–75

Hard rubber troll nodder. "World's Best
Roommate". Marked "Hong Kong
Berries". 4". $50–75

Hard rubber troll nodder man. Marked
"Heico". 6". $50–75

Hard rubber troll nodder woman. Marked
"Western Germany". Strega. 9". $50–65

Hard rubber troll nodder man. Marked
"Heinz and Co." 1967. 5". $60–75

Hard rubber troll nodder man. Marked
"Heico". 9". $60–85

Hard rubber nodder troll, German hunchback. Marked "Heico". 9". $75–100

Hard rubber troll nodders. Marked "Germany". Heico. 9". $5–100 and 7". $50–75

Hard rubber nodder troll. Marked "Germany". Heico. 5". $50–75

Hard rubber lady troll nodder. Marked "Heico Germany". 8". $150–200

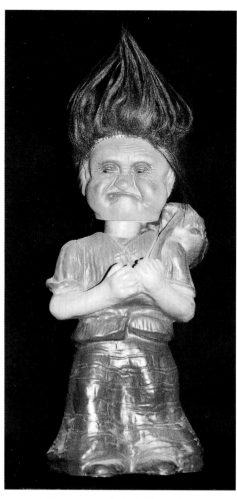

Hard rubber lady troll nodder.
Marked "Heico". 9". $60–75

Hard rubber pirate troll nodder.
8". Unmarked. $40–50

Rubber Irish troll nodder.
Marked "Heico". 8". $50–75

Hard rubber nodder troll on skis.
7.5". Unmarked. $40–50

Hard plastic nodder troll. Marked
"Western Germany Heico". 9". $60–75

Hard rubber Irish troll nodder.
Marked "Heico". 9". $50–75

Hard rubber monk troll nodder. Marked "Western Germany Heico". 9". $50–75

Hard rubber hiker troll. Marked "Germany Heico". 9.5". $50–75

Plastic troll with fur cuffs and collar, nodder. No marks. 7". $150–200

Plastic red-haired Troll nodder. Large plastic nose. 8". Rare. $100–150

Plastic pipe-smoking nodder troll. Rare. (Note large nose). Missing stick in hand. 4.5". $75–100

Plastic pipe-smoking nodder troll. Rare. (Note large nose). 4.5". $75–100

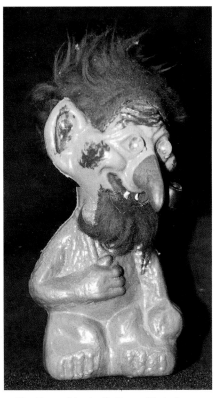

Plastic nodder troll. Rare. (Note large nose). 4.5". $75–100

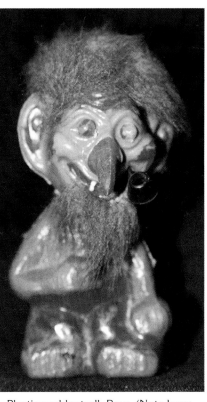

Plastic nodder troll. Rare. (Note large nose). 4.5". $75–100

Papier maché "Lucky Shnook" nodder. Japan. $35–45

Papier maché "Lucky Nik" nodder. 5". Japan. $25–35

Papier mache basketball troll nodder. 6". $30–45

Hard rubber Russ troll nodder. N.Y. Yankees. 1990s. $20–30

Animals

Rubber "Rudy" the elephant troll.
Norfin's Ark. 1989. 3". $7–10

Rubber "Cecily" mouse troll.
3". Norfin's Ark. $7–10

Rubber "KoKo" monkey troll.
3". Norfin's Ark. $7–10

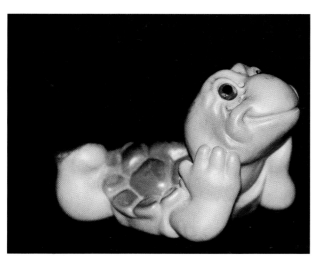

Rubber turtle troll from Dam Things
Designs. 1960s. 4" x 6". $75–100

Rubber monkey troll.
Marked "USA". 3". $40–50

Rubber "Uglies" lion troll. Made in Japan. 3". $30–40 with tag

Rubber "Wiley" fox Troll. Had tag with "International Passport". No Marks. Leprechaun Ltd. Dublin, Ireland. 1970. $75–100

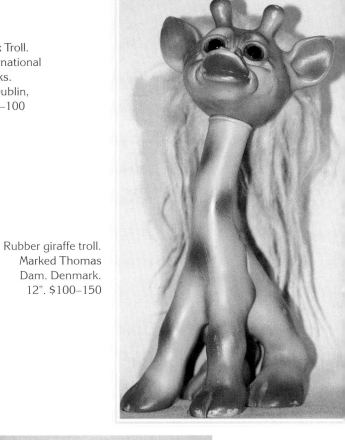

Rubber giraffe troll. Marked Thomas Dam. Denmark. 12". $100–150

Hard plastic Dam Things lion troll. 5". $75–125

Rubber cow troll Dam Things. 3". $30–40

Rubber cow troll. Limited Edition,
Thomas Dam. Denmark. 7". $75–100

Plastic horse troll. Hong Kong. 5". $30–40

Rubber horse troll. Plastic inserted
eyes. Unmarked. 4". $20–30

Hard plastic lion troll. 5".
Unmarked. $30–40

Heavy textured donkey troll. Marked "Dam
Things Est". 9". $200–250

Rubber "Uglies" elephant troll. Marked "Japan" on bottom of belly. 3.5". $25–35

Rubber elephant and donkey troll finger puppets. Dam Things. 3". $35–45

Rubber elephant finger puppet troll. Dam Things. 2.5". $35–45

Rubber elephant troll. Japan. 4" x 5". $25–35

Plastic animal troll. Fur trimmed. Unmarked. 2". $15–20

Hard rubber monkey troll. 7".
Thomas Dam. Rare. $300–400

Rubber DinoTrolls. The series also features 6
Clans: Emo, Glug, Tog, Ega, Uma, Oota.
Original set. 1992. My Kids Toy Manufacturing
Co. Hong Kong.

Rubber fur-trimmed "Dolly"
mouse troll. 3.5". $15–20

Rubber fur-trimmed
cat. 3". $15–20

Rubber DinoTrolls. Cave man, woman and child. Hard rubber. 7". Marked "My Kids Toy Manufacturing Co". 1992. Came as a set. $6–8 each or $30–40 set

Rubber DinoTrolls dinosaurs. Made in China. Marked "My Kids Toy Mf'g Co." $6–8 each

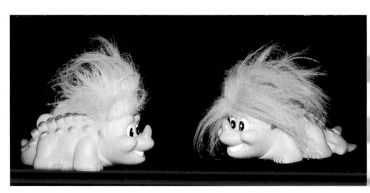

Rubber dinosaur trolls. Trendmasters. $6–8 each

Rubber dinosaur troll. Marked "My Kids Toy Mf'g Co." $6–8

Hard rubber dinosaur. Playskool. 1986. 8". $5–7

Vinyl posable bear troll. 5.5". $5–7

Plastic monster trolls. Trendmasters. Made in Japan. Press belly and they talk. "Give me a kiss" or "You're my best friend". 4". $8–12 each

Rubber horse troll. Marked "Dam Things". 3" x 2.5". $35–45

Rubber key chain, lion troll. 2". $4–5

Rubber troll earrings. Wishing Star. 1". $3–4 each

Rubber earrings. 3". $3–4

Plastic Santa troll earrings and pin. Russ. 1". $3–4 set

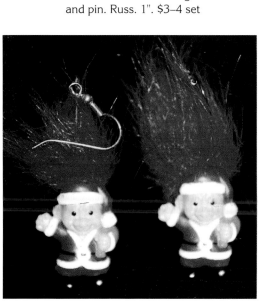

Plastic Santa troll earrings. Russ. $3–4 pair

Plastic ring. $3–4

Plastic troll rings. 1". $3–4 each

Rubber Graduate Troll pin. Russ. 3". $6–8

Goldtone metal necklace and pins. Missing eyes. 2". $10–15 each

Metal troll button. Hair comes out of the back of the button. Norfin. Dam Designs. 2". $5–10 each

Goldtone metal pin with movable eyes. 2". $10–15

Tin pin-back buttons. 3". $5–10 each

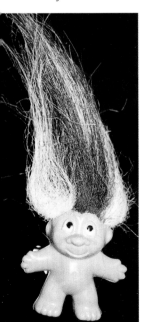

Rubber troll pin. Norfin. $5–8

Plastic troll watch. Russ. Early 1990s. $10–15

Rubber Irish pin. 1". $6–8

Plastic Troll–Y–Bell necklace. Korea. New Creative Enterprises. $3–5

Rubber troll necklace. China. 3". $3–4

Rubber troll necklace. 1". $3–4

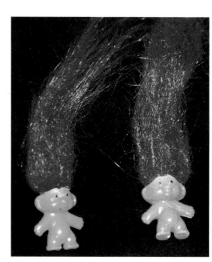

Plastic charms. China. 1.5". $2–3

Plastic charm—angel. Russ. 1". $2–3

Plastic necklace. 2.5". $3–4

Plastic charms. China. 2". $2–3

Rubber troll key ring. 1.5". $3–4

Plastic charms. 1". $2–3 each

Rubber "Total Trollers" key ring. Marked "Street Kids Corp." 1991. 3". $3–4

Rubber Troll key ring. 1989. Dam. 1.5". $3–5

Plastic key rings. Early. .05". $5–8 each

Plastic key ring. Russ. 2.5". $3–4

Rubber "Total Trollers" key ring. Marked "Street Kids Corp." China. $3–4

Rubber "Total Trollers" key ring. Marked "Street Kids Corp." China. $3–4

Rubber trollette key ring. Korea. 2.5". $3–4

Plastic Christmas troll key ring. 3". $3–4

Rubber vending machine prize key ring. 1". $2–3

Minis and Pencil Toppers

Rubber troll pencil toppers. $3–5 each

Rubber troll baby pencil topper. Applause. 1990s. $4–8

Rubber troll pencil toppers. 1.5".
Marked "S.H.E." $3–5 each

Rubber troll pencil toppers. 1.5".
Marked "S.H.E." $3–5 each

Rubber troll pencil toppers. Russ. $3–5 each

Hard plastic troll pencil toppers. China. 1.5". $3–5

Plastic and rubber troll pencil toppers. 3/4"–1.5". $3–5

Plastic troll
pencil topper.
1.5". $3–5

Plastic troll
pencil topper.
1.5". $3–5

Plastic troll pencil topper. 1". $3–5

Plastic green troll
pencil topper. China.
1.5". $3–5

Plastic troll pencil topper. 1.5". $3–5

Rubber troll
pencil topper.
Russ. 1.5". $4–5

Rubber mini trolls. 1.5". $3–5

Plastic Thanksgiving mini
troll. Russ. 1". $3–4

Mini troll figure. Ace novelty.
1.5". $3–5

Rubber mini trolls. Russ. 3.5". $3–5 each

Paper, Books and Puzzles

Norfin memo pad.
3 x 4.5". $4–5

Note pad. 4 x 5". China. $4–5

Paper bag. Russ
Berrie and Co.
7". $3–4

Paper napkins.
Treasure Trolls. $2–3

Troll wrapping paper. $3–4 sheet

Troll trading cards. Norfin. In and out of package. $2–3 each

McCall patterns. 1964. $8–12 each

Stickers from Hallmark. $5–7

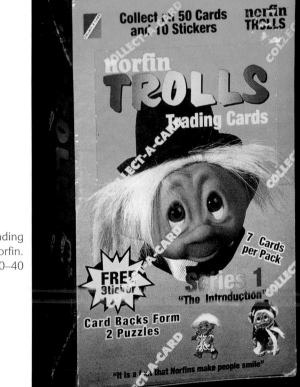

Box of trading cards. Norfin. 1992. $30–40

Scholastic magazine, *Hot Dog*. 1992. $4–5

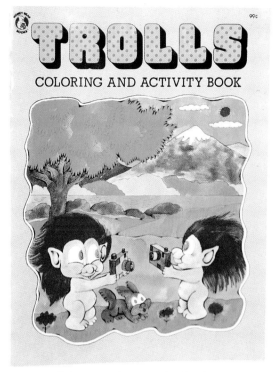

Coloring and activity book. Honey Bear. 1990s. $3–4

Norfin troll book. *Ralph Troll's New Bicycle*. 1992. $3–4

Laugh Out Loud joke book. Norfin Trolls. 1992. $3–4

Troll Jokes and Riddles. Watermill Press. 1993. $5–7

Wishnik Cut-Outs. Uneeda Toy Co. 1965.
Western Press. $20–25

Cardboard puzzle. Russ. Golden.
8" x 11". $4–6

Wishnik Cut-Outs. Uneeda Toy
Co. 1965. $20–25

Cardboard puzzle. Russ. Golden.
12" x 14". 1990s. $4–6

Puppets and Toys

Hand puppet,
rubber head.
$30–40

Magician troll
hand puppet,
hard rubber head.
Russ. 11". $10–15

Chef troll hand
puppet, rubber
head. Russ. $10–15

Doctor hand puppet, rubber
head. $10–15

Fireman hand puppet,
rubber head. Russ.
Thailand. $10–15

Plastic bear troll on scooter. $4–6

Plastic lady troll. Wind-up walker. China. $5–7

Rubber finger puppet. Dapp. China. $5–8

Plastic bunny troll. Wind-up walker. Russ. $5–7

Plastic troll. Wind-up walker. China. $5–7

Rubber finger puppet. Phillipines. 3". $10–15

Plastic battery operated skateboarder troll.
Russ. 9" x 10". $10–15

Plastic skateboard troll. Jessica. 5". $6–8

Plastic surfer troll. 4". $8–10

Plastic troll in fins. $6–8

Rubber troll on plastic
scooter. 3" x 3". $8–10

Rubber troll on plastic scooter. $8–10

Rubber troll on plastic skateboard. $6–10

Plastic troll on friction skateboard. Soma, China. 1966. 3". $10–15

Plastic train and fire truck friction toys. Russ. 2"x 2". Early 1990s. $5–8 each

Plastic wind-up car. Russ. 2.5". $5–8

Front and back of Treasure Trolls box showing all items in this series. Teeter Totter. Plastic. Ace Novelty. 1990s. $10–15 each

"Trolls on the Go" friction toy. 3" x 4". Russ. $10–15

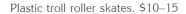

Plastic troll roller skates. $10–15

Plastic troll tape player with headphones. $12–15

121

Plastic push-
up puppet.
British patent.
Late 1980s.
3". $6–10

Plastic bicycle handle. A mint-
condition pair of these would be
$60–75. Rare. As is $10–15

Magic Trolls and the Troll Warriors
video by Applause. 1991. $7–10

Skater, rubber head,
plastic body. Battery
operated. Leader Toy
Corp. $30–40

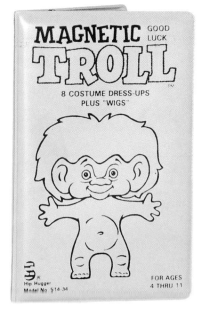

Troll playing cards. Norfin. $5–7

Vinyl Magnetic Troll board.
Magnetic Troll and clothes Inside.
Smethport, Pa. Specialty Co. Hip
Hugger. 4" x 6". $10–15

Search for the Wishstones Game.
Pressman Co. Treasure Trolls. $8–10

Cloth stuffed color-me doll.
ITB. International Bon Ton
Toys. 1990. 9". $8–12

Plastic yoyo.
Hong Kong.
1.5". $4–6

Plastic
kaliedoscope.
Russ. 1.5".
$4–6

Rubber poseable Barbie. She is
wearing a nylon troll print outfit
with troll earrings. Mattel Inc; 1976
marked on back of neck. $20–30

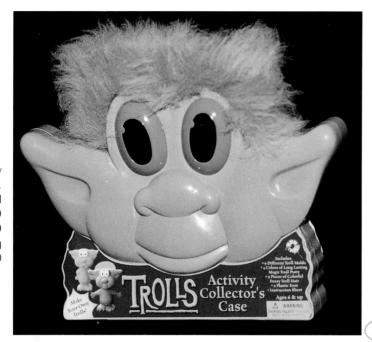

Plastic troll Activity Collectors Case. Includes 9 different Troll molds, 4 putty colors, 9 pieces of fuzzy hair, 9 eyes. Made by Flying Colors. 1990s. $15–20

Inside and outside of vinyl troll house. Ideal Toy Corp. $35–50

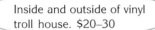

Inside and outside of vinyl troll house. $20–30

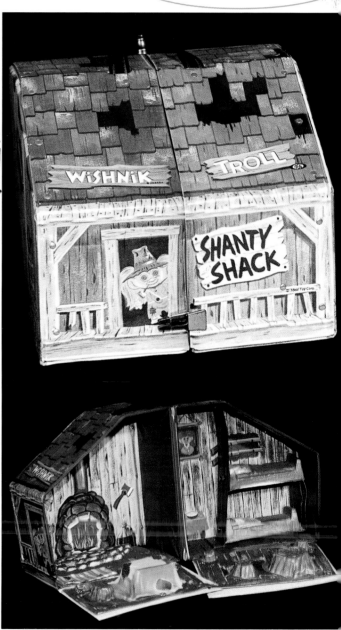

Life in the Troll House

Wishniks vinyl Nik house. Ideal Toy Corp. $20–40

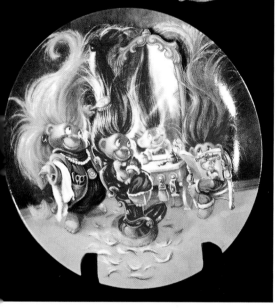

Ceramic Troll plate. Monday's Troll by Tom Dubois. Crestley Collections. 1993. 8". $20–30

Ceramic troll plate. 8.5". Ace Novelty Co. $15–20

Ceramic troll plate. 8.5". Ace Novelty Co. $15–20

Ceramic troll plate. From original painting by Chris Hopkins. "Enchanted Norfin Trolls". Reco. 1994. 6.5". $30–40

Plastic Trollies triplets tumbler.
Peter Pan Inc. 8oz. $3–4

Rubber stamp. 2". $3–4

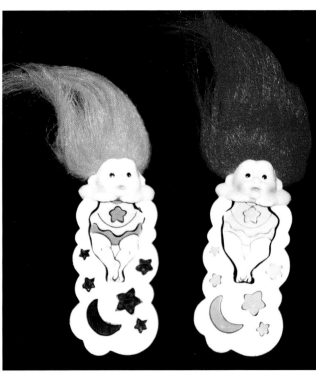

Plastic troll bookmarks. Applause. 3.5". $3–4 each

Rubber stamp. 2". $4–5

Metal troll tray. 12" x 17". $5–8

Russ troll outfit. $5–8

Rubber troll heads on shampoo bottles. Norfin. $5–8

Rubber troll sitting on top of cologne bottle. Norfin. The Troll Co. 1992. $7–10

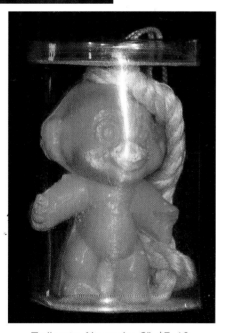

Troll soap. No marks. 3". $7–10

Plastic troll combs. 2" x 2". $2–3

Plastic troll on elastic belt. Norfin. 1.5" Troll. $6–8

Plastic troll on ribbon. Korea. $4–5

Plastic troll on elastic belt. Norfin. 1.5" Troll. $6–8

Soft rubber troll on elastic suspenders. 3" Troll. $6–8

Plastic troll shoe ties. 1". $3–5

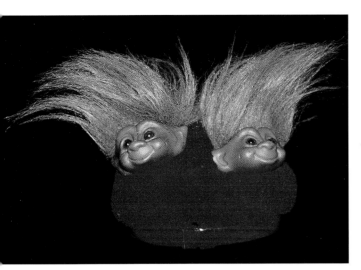

Rubber and plush troll earmuffs. $8–12

Cotton troll sheet set. Treasure Trolls. $5–7 set

Child's slippers. Plush with rubber troll head. $7–10

Cotton stuffed pillow doll. No marks. 18". $10–15

Child's slippers. Plush with rubber troll head. Norfin. $7–10

Fabric troll pillows. 12". Ace Novelty co. $6–10 each

Plastic toothbrush holder. Norfin. Missing toothbrush. $10–15 as is

Supertroll plastic water bottle. $3–4

Plastic troll purse. Sparkles inside. 6" tall. $5–8

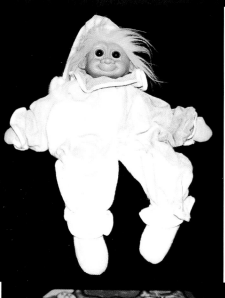

Clown zip-up pajama bag. 22". 1990s. $12–15

Vinyl lunch tote. Norfin. $10–15

Troll House cookie tin. SCC China. 6" x 5". $3–4

Treasure Trolls plastic lunch box. Alladin. $10–15

Vinyl backpack. Norfin. $10–15

Plastic troll carrier. Ace Novelty Co. 1992. $15–20

Plastic backpack. Norfin. $10–15

Plastic pencil box. Ace Novelty Co. $5–6

Plush Christmas stocking with rubber-head Santa troll. Marked "Mithy China". $20–30 new

Vinyl bag. Ace Novelty Co. $6–8

Plastic plug-in night light. China. $5–8

Rubber Viking. No marks. Poor condition. Missing beard and hair. 6.5". $20–30 as is

Rubber troll with green eyes. No marks. 3". $6–10

Rubber troll girl with red hair. No marks. 3". $4–6

Rubber nurse troll. 6". No marks. $6–10

Plastic genie troll. 3". $7–10

Rubber hippie troll. Unmarked. 3". $20–25

Rubber troll with black hair. Unmarked. 2.5". $7–10

Rubber troll girl in black dress. Unmarked. 3". $7–10

Rubber red-head troll with gunbelt and gun. 3". $7–10

Rubber German troll. Painted-on suit. No marks. 3". $5–20

Rubber troll with movable head. Painted-on suit. Unmarked. 5.5". $10–15

Rubber troll with kitten. Unmarked. 3". $7–10

Rubber nude troll. One green eye and one red eye. 5.5". Unmarked. $15–20

Rubber mini troll. Unmarked. 2.5". $4–5

Soft rubber troll. Unmarked.2.5". $4–5

Plastic mini troll. Unmarked. 2.5". $3–4

Plastic mini troll. Unmarked. 2.5". $4–5

Rubber Indian and
papoose trolls. 3".
Unmarked. $10–15

Rubber collegiate troll. 2.5".
Unmarked. $15–20

Rubber collegiate
trolls. 2.5". Un-
marked. $15–20

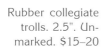

Left:
Rubber
collegiate
troll. 2.5".
Unmarked.
$15–20

Right:
Rubber
collegiate
troll, nude.

Rubber troll girl. 2.5". Unmarked. $15–20

Rubber troll with green eyes. 3". $6–10

Rubber nude troll girls. 3". Unmarked. $15–20

Ceramic troll figurines. 5". $30–40 each